# Simple Explanations of Some Useful Mathematical Identities and Theorems

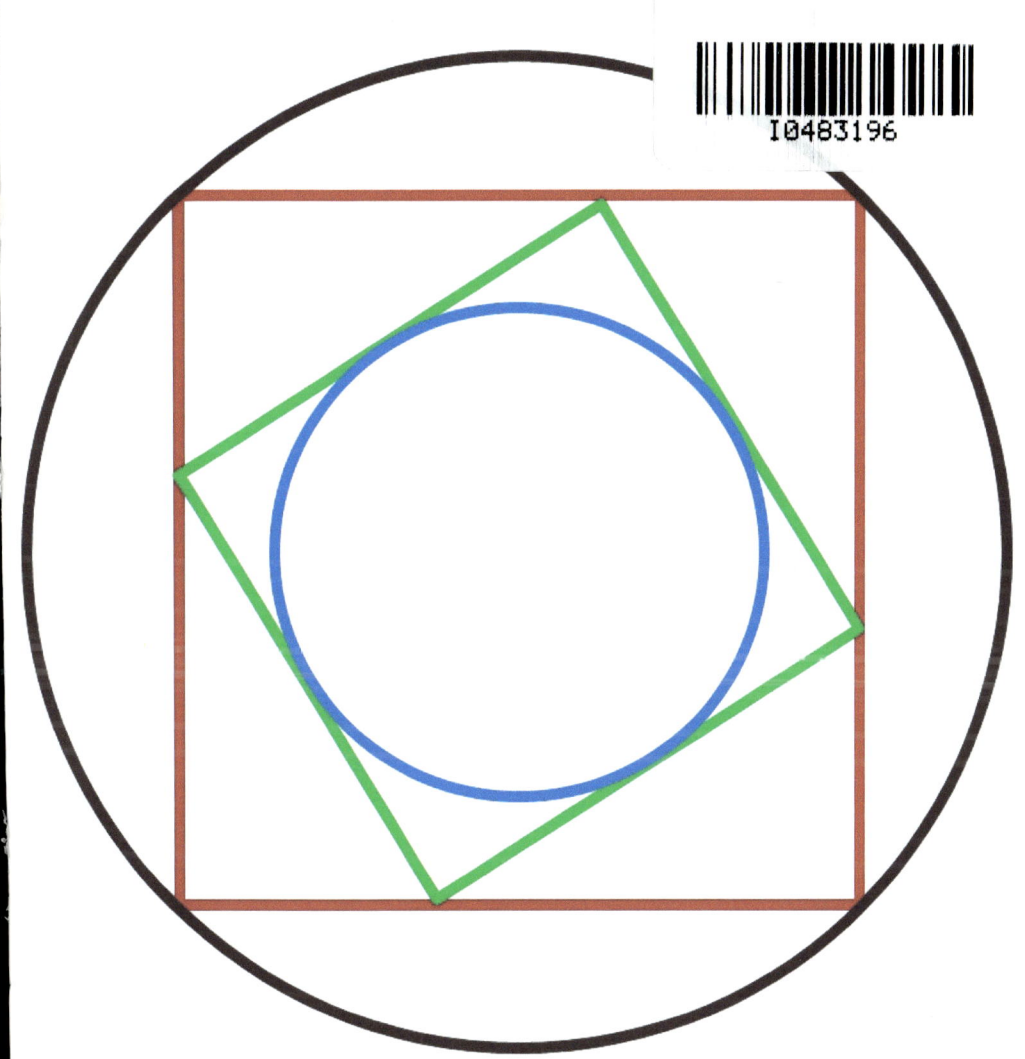

Jeremy Alford, Ph.D

ii

This book is dedicated to my father, Ray Alford. Without his guidance and encouragement when I was young, this would not be possible.

# Contents

# Preface

The goal of this book is to provide simple explanations to some of the most important theorems and identities in mathematics. They are not intended to be rigorous proofs but rather intuitive guides that take the mystery out of some of the most powerful mathematical facts ever discovered. This is not meant to be used as a textbook. It could be used as supplement to formal coursework or by someone who is simply interested in the development of advanced mathematics. The reader is expected to have a working knowledge of the material discussed.

Much of this book deals with theorems in calculus but there is also some algebra, geometry, trigonometry, and number theory. The explanations presented here are the ones the author considers the easiest to understand. They are certainly not the only ones and they may not be the ones used by the people who originally discovered these facts.[1] Historically, the development of mathematics is very long and complicated with similar ideas being developed in different ways by different cultures.[2] Much of the material presented here originated in ancient Greece, Arabia, and $17^{th}$ century Europe. Some of it may be traced to even older cultures, such as, ancient China and Mesopotamia.

This book is the result of the author's attempt to gain a deeper understanding of mathematics. No references are given because the facts presented are well-known and can be verified by many sources.[3] The explanations are the author's independent work and any similarity to work found elsewhere is a result of mathematics being discovered and not invented. Although much effort was used to make the explanations as clear as possible, nothing can replace the student-teacher relationship which allows for questions with immediate feedback and fosters the learning experience.

---

[1] The book, *The Pythagorean Proposition*, Second Edition (1940) by Elisha Scott Loomis contains 370 proofs of the Pythagorean Theorem.

[2] There are many books on the historical development of mathematics including *A History of Mathematics*, Second Edition (1991) by Carl B. Boyer and Uta C. Merzbach.

[3] A useful reference that contains many of the identities and theorems presented here is *Schaum's Outlines Mathematical Handbook of Formulas and Tables*, Second Edition (1999) by Murray R. Spiegel, Ph.D and John Liu, Ph. D.

# 1 The Pythagorean Theorem

---

*The Pythagorean Theorem*

Any right triangle with legs of length $A$ and $B$ and hypotenuse of length $C$ satisfies

$$A^2 + B^2 = C^2 \tag{1}$$

---

The importance of this theorem cannot be understated, however, the proof is very simple. Consider the two squares shown in Fig.1. Here, we have a square of side-length $C$ inside another square of side-length $A + B$. The area of the outer square can be calculated in two ways. The first way is to simply square the length of one side, the usual way of finding the area of the square. The second way is to note that the outer square is composed of five smaller figures, the inner square and four triangles. The area of the inner square is $C^2$ and the area of each triangle is $\frac{1}{2}AB$. The area of the outer square must be equal to the sum of the areas of these five figures. These two methods for obtaining the area of the outer square are summarized below,

$$area \;=\; (A+B)^2 \;, \tag{2}$$
$$area \;=\; C^2 + 4\left(\frac{1}{2}AB\right) \;. \tag{3}$$

Since these two expressions must be equal to each other, we have,

$$(A+B)^2 = C^2 + 4\left(\frac{1}{2}AB\right) \;. \tag{4}$$

Simplifying the above equation gives,

$$A^2 + B^2 = C^2 \; , \tag{5}$$

which is the Pythagorean Theorem.

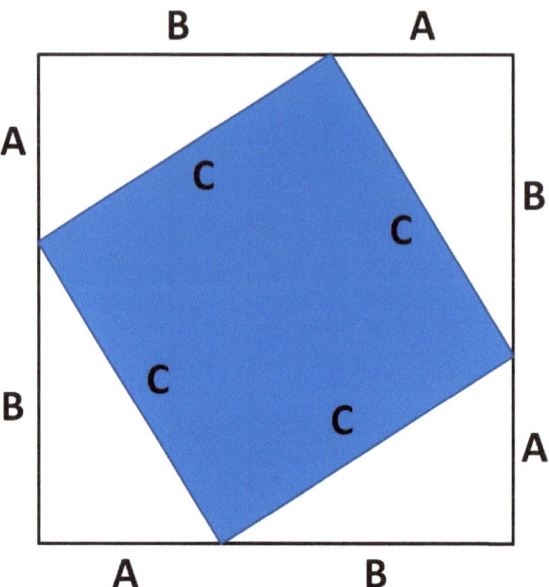

Figure 1: The area of the inner square is $C^2$ and the area of the outer square is $A + B)^2$. The area of the outer square can also be obtained by adding the area of the inner square and the areas of the four triangles. Requiring these two expressions for the same area to be equal leads to $A^2 + B^2 = C^2$.

# 2  Trigonometry

## 2.1  The Fundamental Trigonometric Identity

---

*The Fundamental Trigonometric Identity*

For any angle $\theta$,

$$\sin^2 \theta + \cos^2 \theta = 1 \tag{6}$$

---

To prove this identity, we must first use the Pythagorean Theorem to derive the equation for a circle. Consider the circle shown in Fig.2. A right triangle can be formed with one vertex at the origin and another vertex at any point on the circle. Since the Pythagorean Theorem must be true for any right triangle drawn on a flat surface, we can apply it to any triangle formed in the way described in the figure. The lengths of the legs of the triangles are given by the coordinates of the points on the circle. The figure shows a triangle formed from a point where both coordinates are positive but this is not necessary since the coordinates will be squared when the Pythagorean Theorem is applied. Applying the Pythagorean Theorem gives,

$$x^2 + y^2 = r^2 \ , \tag{7}$$

which is the equation for a circle of radius $r$ and centered at the origin in the $xy$ plane.

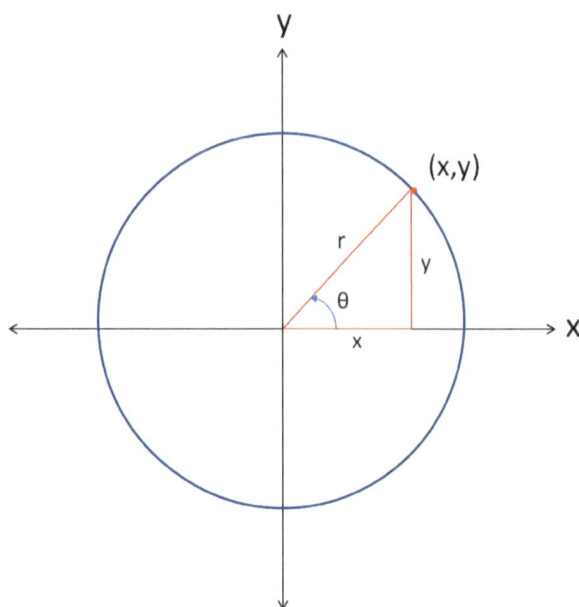

Figure 2: A circle of radius $r$ is centered at the origin in the $xy$ plane. A right triangle can be formed by the $x$-axis, a line segment along a radius, and a line segment, parallel to the $y$-axis, from any point on the circle to the $x$-axis. The hypotenuse of this triangle has length $r$ and the legs have lengths $|x|$ and $|y|$. Applying the Pythagorean Theorem gives the equation for the circle, $x^2 + y^2 = r^2$.

The trigonometric functions can be defined in terms of the coordinates of the circle in Fig.2 by the equations,

$$\sin \theta = \frac{y}{r} , \tag{8}$$

$$\cos \theta = \frac{x}{r} , \tag{9}$$

where $\theta$ is angle formed by the positive $x$-axis and the line segment connecting the origin and the point $(x, y)$ on the circle, measured in the counter-clockwise direction, as shown in the figure. Solving the above equations for $x$ and $y$ and substituting them into Eq.7 gives,

$$(r \cos \theta)^2 + (r \sin \theta)^2 = r^2 . \tag{10}$$

Now, dividing both sides of the above equation by $r^2$, we have,

$$(\cos\theta)^2 + (\sin\theta)^2 = 1 , \tag{11}$$

or

$$\sin^2\theta + \cos^2\theta = 1 . \tag{12}$$

which is the Fundamental Trigonometric Identity.

## 2.2  Addition Formulas

*Addition Formulas*

$$\sin(\theta_1 \pm \theta_2) = \sin\theta_1 \cos\theta_2 \pm \cos\theta_1 \sin\theta_2 \tag{13}$$

$$\cos(\theta_1 \pm \theta_2) = \cos\theta_1 \cos\theta_2 \mp \sin\theta_1 \sin\theta_2 \tag{14}$$

$$\tan(\theta_1 \pm \theta_2) = \frac{\tan\theta_1 \pm \tan\theta_2}{1 \mp \tan\theta_1 \tan\theta_2} \tag{15}$$

We can prove the first two of these identities by referring to Fig.3 which shows two coordinate systems rotated by an angle of $\phi$ with respect to each other. The line segment connecting the origins to a point on the plane is at an angle of $\theta$ with respect to the $x$-axis and an angle of $\theta'$ with respect to the $x'$-axis. The three angles are related to each other by the equation,

$$\theta = \theta' + \phi . \tag{16}$$

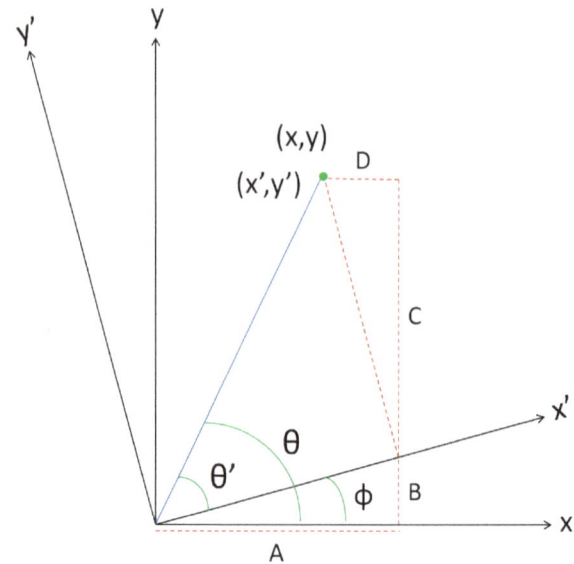

Figure 3: The $x'y'$ coordinates are rotated by an angle of $\phi$ relative to the $xy$ coordinates. The length of the line segment connecting the origins to a point on the plane is the same in each coordinate system.

The hypotenuse of the triangle formed by the dashed lines has length $y'$ and the angle formed by the hypotenuse and the side labeled $C$ is the same as $\phi$ (to see this, imagine rotating the triangle clockwise by 90 degrees). The lengths of the other four dashed lines, labeled $A$, $B$, $C$, and $D$, are related to the lengths $x$ and $y$ according to,

$$x = A - D \tag{17}$$

$$y = B + C . \tag{18}$$

For simplicity, let's assume the length of the solid line connecting the origin to the point $(x, y)$ (or $(x', y')$, depending on which coordinate system we are looking at) is equal to one.

From basic trigonometry, we know,

$$A = x' \cos \phi \tag{19}$$

$$B = x' \sin \phi \tag{20}$$

$$C = y' \cos \phi \tag{21}$$

$$D = y' \sin \phi . \tag{22}$$

We also know,

$$x = \cos \theta \tag{23}$$

$$y = \sin \theta \tag{24}$$

$$x' = \cos \theta' \tag{25}$$

$$y' = \sin \theta' . \tag{26}$$

Combining Eq.18 - Eq.26 gives,

$$\cos \theta = \cos \theta' \cos \phi - \sin \theta' \sin \phi \tag{27}$$

$$\sin \theta = \cos \theta' \sin \phi + \sin \theta' \cos \phi . \tag{28}$$

Now, using Eq.16, we have,

$$\cos(\theta' + \phi) = \cos \theta' \cos \phi - \sin \theta' \sin \phi \tag{29}$$

$$\sin(\theta' + \phi) = \cos \theta' \sin \phi + \sin \theta' \cos \phi . \tag{30}$$

To get the addition formulas for $\sin(\theta' - \phi)$ and $\cos(\theta' - \phi)$ we simply replace $\phi$ with $-\phi$ and use the facts that $\cos \phi$ is an even function ($\cos(-\phi) = \cos \phi$) and $\sin \phi$ is an odd function

$$(\sin(-\phi) = -\sin\phi),$$

$$\cos(\theta' - \phi) = \cos\theta' \cos\phi + \sin\theta' \sin\phi \tag{31}$$

$$\sin(\theta' - \phi) = -\cos\theta' \sin\phi + \sin\theta' \cos\phi. \tag{32}$$

Changing the labels of the angles according to,

$$\theta' \rightarrow \theta_1 \tag{33}$$

$$\phi \rightarrow \theta_2, \tag{34}$$

and rearranging the terms on the right side of Eq.30 and Eq.32 gives the formulas for $\sin(\theta_1 \pm \theta_2)$ and $\cos(\theta_1 \pm \theta_2)$ as listed at the beginning of this section.

To get the formula for $\tan(\theta_1 \pm \theta_2)$ we use,

$$\tan\theta = \frac{\sin\theta}{\cos\theta}. \tag{35}$$

$$\tan(\theta_1 \pm \theta_2) = \frac{\sin(\theta_1 \pm \theta_2)}{\cos(\theta_1 \pm \theta_2)}$$

$$= \frac{\sin\theta_1 \cos\theta_2 \pm \cos\theta_1 \sin\theta_2}{\cos\theta_1 \cos\theta_2 \mp \sin\theta_1 \sin\theta_2}. \tag{36}$$

Dividing the numerator and denominator of the right side of the above equation by $\cos\theta_1 \cos\theta_2$ and using Eq.35 gives the final result,

$$\tan(\theta_1 \pm \theta_2) = \frac{\tan\theta_1 \pm \tan\theta_2}{1 \mp \tan\theta_1 \tan\theta_2}. \tag{37}$$

Other useful formulas, such as multiple angle and half-angle formulas, can easily be derived using the addition formulas given above and the Fundamental Trigonometric Identity (Eq.6).

## 2.3  Sum and Difference Formulas

*Sum and Difference Formulas*

$$\sin\theta_1 + \sin\theta_2 \;=\; 2\sin\left(\frac{1}{2}(\theta_1+\theta_2)\right)\cos\left(\frac{1}{2}(\theta_1-\theta_2)\right) \tag{38}$$

$$\sin\theta_1 - \sin\theta_2 \;=\; 2\cos\left(\frac{1}{2}(\theta_1+\theta_2)\right)\sin\left(\frac{1}{2}(\theta_1-\theta_2)\right) \tag{39}$$

$$\cos\theta_1 + \cos\theta_2 \;=\; 2\cos\left(\frac{1}{2}(\theta_1+\theta_2)\right)\cos\left(\frac{1}{2}(\theta_1-\theta_2)\right) \tag{40}$$

$$\cos\theta_1 - \cos\theta_2 \;=\; 2\sin\left(\frac{1}{2}(\theta_1+\theta_2)\right)\sin\left(\frac{1}{2}(\theta_2-\theta_1)\right) \tag{41}$$

To prove these identities we will refer to Fig.4. The coordinates of the points $(x_1, y_1)$ and $(x_2, y_2)$ are related to the angles, $\theta_1$ and $\theta_2$, by the equations.

$$x_1 \;=\; \cos\theta_1\,, \tag{42}$$

$$y_1 \;=\; \sin\theta_1\,, \tag{43}$$

$$x_2 \;=\; \cos\theta_2\,, \tag{44}$$

$$y_2 \;=\; \sin\theta_2\,. \tag{45}$$

The coordinates, $x$ and $y$, can be expressed as,

$$x \;=\; L\cos\left(\frac{1}{2}(\theta_1+\theta_2)\right)\,, \tag{46}$$

$$y \;=\; L\sin\left(\frac{1}{2}(\theta_1+\theta_2)\right)\,, \tag{47}$$

where $L$ is the length of the line segment connecting the origin and the point $(x, y)$ (dashed red line). Using the fact that the definition of a tangent line requires it to be perpendicular to the line containing the center of the circle and the point of tangency, we can use basic trigonometry to relate $L$ to the angles, $\theta_1$ and $\theta_2$,

$$\cos\left(\frac{1}{2}\left(\theta_1 - \theta_2\right)\right) = \frac{1}{L}. \tag{48}$$

We now have,

$$x = \frac{\cos\left(\frac{1}{2}\left(\theta_1 + \theta_2\right)\right)}{\cos\left(\frac{1}{2}\left(\theta_1 - \theta_2\right)\right)}, \tag{49}$$

$$y = \frac{\sin\left(\frac{1}{2}\left(\theta_1 + \theta_2\right)\right)}{\cos\left(\frac{1}{2}\left(\theta_1 - \theta_2\right)\right)}. \tag{50}$$

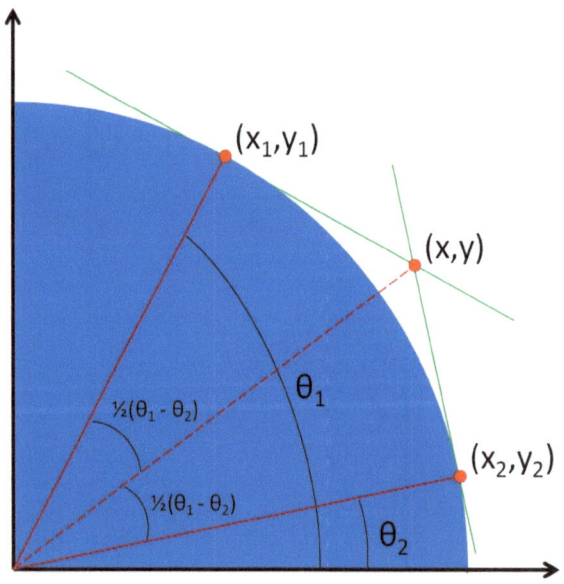

Figure 4: Unit quarter-circle centered at the origin. The green lines are tangent to the circle at points $(x_1, y_1)$ and $(x_2, y_2)$ and intersect each other at the point $(x, y)$. The dashed red line bisects the angle formed by the solid red lines.

The slope of the line containing the origin and the point $(x_1, y_1)$ (upper red line) is $\frac{y_1}{x_1}$ and the slope of the line containing the origin and the point $(x_2, y_2)$ (lower red line) is $\frac{y_2}{x_2}$. The slope the line tangent to the circle at the point $(x_1, y_1)$ (upper green line) is $-\frac{x_1}{y_1}$ and the slope the line tangent to the circle at the point $(x_2, y_2)$ (lower green line) is $-\frac{x_2}{y_2}$.[4] Both tangent lines contain the point $(x, y)$ so we can also calculate the slopes using the differences in the coordinates. This gives,

$$-\frac{x_1}{y_1} = \frac{y - y_1}{x - x_1} , \tag{51}$$

and

$$-\frac{x_2}{y_2} = \frac{y - y_2}{x - x_2} . \tag{52}$$

Rearranging these equations and using the Fundamental Trigonometric Identity (Eq.6), we have,

$$x_1 x + y_1 y = 1 , \tag{53}$$

and

$$x_2 x + y_2 y = 1 . \tag{54}$$

Solving these equations for $x$ and $y$ gives,

$$x = \frac{y_1 - y_2}{x_2 y_1 - x_1 y_2} , \tag{55}$$

$$y = \frac{x_1 - x_2}{x_1 y_2 - x_2 y_1} . \tag{56}$$

---

[4]The slope of a line tangent to a circle is the negative reciprocal of the line containing the center of the circle and the point of tangency.

Using Eq.42 - Eq.45 and the first addition formula from Sec.2.2 (Eq.13), the denominators can be simplified,

$$x_2 y_1 - x_1 y_2 = \sin(\theta_1 - \theta_2) , \tag{57}$$

$$x_1 y_2 - x_2 y_1 = \sin(\theta_2 - \theta_1) . \tag{58}$$

We now have,

$$x = \frac{\sin\theta_1 - \sin\theta_2}{\sin(\theta_1 - \theta_2)} , \tag{59}$$

$$y = \frac{\cos\theta_1 - \cos\theta_2}{\sin(\theta_2 - \theta_1)} , \tag{60}$$

where the numerators have been rewritten using Eq.42 - Eq.45. Rearranging these equations and expressing $x$ and $y$ using Eq.49 and Eq.50, gives,

$$\sin\theta_1 - \sin\theta_2 = \frac{\cos\left(\frac{1}{2}(\theta_1 + \theta_2)\right)}{\cos\left(\frac{1}{2}(\theta_1 - \theta_2)\right)} \sin(\theta_1 - \theta_2) , \tag{61}$$

$$\cos\theta_1 - \cos\theta_2 = \frac{\sin\left(\frac{1}{2}(\theta_1 + \theta_2)\right)}{\cos\left(\frac{1}{2}(\theta_1 - \theta_2)\right)} \sin(\theta_2 - \theta_1) . \tag{62}$$

Using the Double Angle Formula, $\sin(2\theta) = 2\sin\theta\cos\theta$,[5] We can write,

$$\frac{\sin(\theta_1 - \theta_2)}{\cos\left(\frac{1}{2}(\theta_1 - \theta_2)\right)} = 2\sin\left(\frac{1}{2}(\theta_1 - \theta_2)\right) , \tag{63}$$

$$\frac{\sin(\theta_2 - \theta_1)}{\cos\left(\frac{1}{2}(\theta_1 - \theta_2)\right)} = 2\sin\left(\frac{1}{2}(\theta_2 - \theta_1)\right) , \tag{64}$$

where we have used the fact that cosine is an even function, $\cos\left(\frac{1}{2}(\theta_1 - \theta_2)\right) = \cos\left(\frac{1}{2}(\theta_2 - \theta_1)\right)$,

---

[5] The Double Angle Formula can easily be proved with the Addition Formula from Sec.2.2 (Eq.13) by simply setting $\theta_1 = \theta_2$.

in the second equation. We can now write Eq.61 and Eq.62 as,

$$\sin \theta_1 - \sin \theta_2 \;=\; 2\cos\left(\frac{1}{2}(\theta_1 + \theta_2)\right)\sin\left(\frac{1}{2}(\theta_1 - \theta_2)\right) , \qquad (65)$$

$$\cos \theta_1 - \cos \theta_2 \;=\; 2\sin\left(\frac{1}{2}(\theta_1 + \theta_2)\right)\sin\left(\frac{1}{2}(\theta_2 - \theta_1)\right) , \qquad (66)$$

which are the Difference Formulas, Eq.39 and Eq.41.

The Sum Formulas can be easily obtained from the Difference Formulas. By making the substitution, $\theta_2 \rightarrow -\theta_2$, in Eq.39, using the fact that sine is an odd function $(\sin(-\theta) = -\sin\theta)$, and interchanging the terms on the right side of the equation, we have,

$$\sin \theta_1 + \sin \theta_2 \;=\; 2\sin\left(\frac{1}{2}(\theta_1 + \theta_2)\right)\cos\left(\frac{1}{2}(\theta_1 - \theta_2)\right) , \qquad (67)$$

which is the Sum Formula, Eq.38.

By making the substitution, $\theta_2 \rightarrow \theta_2 + 180°$, in Eq.41 and using the facts that adding 180 degrees to the argument of a cosine function changes the sign $(\cos(\theta + 180°) = -\cos\theta)$ and adding 90 degrees to the argument of a sine function changes it to a cosine function $(\sin(\theta + 90°) = \cos\theta)$, we have,

$$\cos \theta_1 + \cos \theta_2 \;=\; 2\cos\left(\frac{1}{2}(\theta_1 + \theta_2)\right)\cos\left(\frac{1}{2}(\theta_1 - \theta_2)\right) , \qquad (68)$$

which is the Sum Formula, Eq.40. We have also used the fact that cosine is an even function.

13

## 2.4 The Law of Sines

*The Law of Sines*

For any angle triangle on a plane,

$$\frac{\sin\theta_A}{A} = \frac{\sin\theta_B}{B} = \frac{\sin\theta_C}{C} \tag{69}$$

where $A$, $B$, and $C$ are the lengths of the sides opposite from the angles with corresponding subscripts (see Fig.5 and Fig.6).

This law is fairly obvious for right triangles where we can define the sine of an angle to be the length of the opposite side divided by the length of the hypotenuse. In this case, each term in Eq.69 is equal to one divided by the length of the hypotenuse. To prove that it is true in general, we will consider acute triangles (all angles are less than 90 degrees) and obtuse triangles (one angle is greater than 90 degrees) separately.

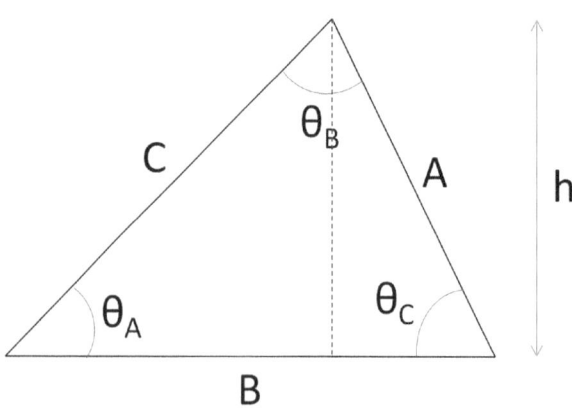

Figure 5: Acute triangle of height $h$.

Consider Fig.5 where we have an acute triangle of height $h$ with sides and angles labeled $A$, $B$, and $C$. From basic trigonometry, we know,

$$\sin \theta_A = \frac{h}{C}, \tag{70}$$

$$\sin \theta_C = \frac{h}{A}. \tag{71}$$

Dividing the first of these equations by $A$ and dividing the second equation by $C$ gives,

$$\frac{\sin \theta_A}{A} = \frac{\sin \theta_C}{C} = \frac{h}{AC}. \tag{72}$$

To show that $\frac{\sin \theta_B}{B}$ is also equal to $\frac{h}{AC}$ will require a little more effort. We start by splitting the angle and side into two parts, along the dashed line, such that,

$$\theta_B = \theta_{B_1} + \theta_{B_2}, \tag{73}$$

and

$$B = B_1 + B_2. \tag{74}$$

Again, using basic trigonometry, we can write,

$$\sin \theta_{B_1} = \frac{B_1}{C}, \tag{75}$$

$$\sin \theta_{B_2} = \frac{B_2}{A}, \tag{76}$$

$$\cos \theta_{B_1} = \frac{h}{C}, \tag{77}$$

$$\cos \theta_{B_2} = \frac{h}{A}. \tag{78}$$

Using these equations and the first addition formula in Sec.2.2 (Eq.13) gives,

$$
\begin{aligned}
\sin(\theta_{B_1} + \theta_{B_2}) &= \sin\theta_{B_1}\cos\theta_{B_2} + \cos\theta_{B_1}\sin\theta_{B_2} \\
&= \frac{B_1}{C}\frac{h}{A} + \frac{h}{C}\frac{B_2}{A} \\
&= \frac{h}{AC}(B_1 + B_2) \ .
\end{aligned}
\tag{79}
$$

Now, dividing both sides of this equation by $B_1 + B_2$ and using Eq.73 - Eq.74 we have,

$$
\frac{\sin\theta_B}{B} = \frac{h}{AC} \ ,
\tag{80}
$$

which finishes the proof of the Law of Sines for acute triangles.

To prove the Law of Sines for obtuse triangles we will refer to Fig.6. Here we have defined two new quantities called $B'$ and $\theta_{C'}$. Because $\theta_C + \theta_{C'} = 180°$, $\sin\theta_{C'}$ and $\cos\theta_{C'}$ are related to $\sin\theta_C$ and $\cos\theta_C$ according to,

$$
\sin\theta_{C'} = \sin\theta_C \ ,
\tag{81}
$$

and,

$$
\cos\theta_{C'} = -\cos\theta_C \ .
\tag{82}
$$

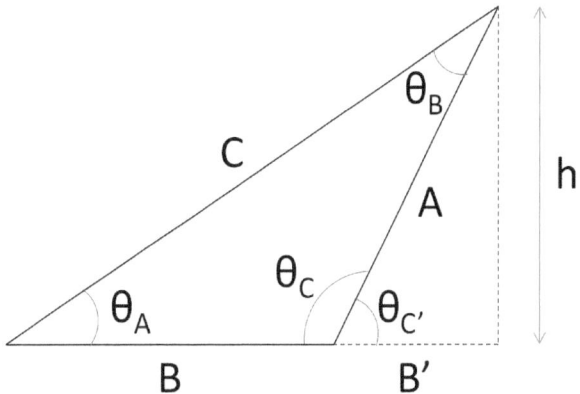

Figure 6: Obtuse triangle of height $h$.

From basic trigonometry, we know,

$$\sin \theta_A = \frac{h}{C}, \tag{83}$$

$$\sin \theta_{C'} = \frac{h}{A}, \tag{84}$$

$$\cos \theta_A = \frac{B + B'}{C}, \tag{85}$$

$$\cos \theta_{C'} = \frac{B'}{A}. \tag{86}$$

Dividing the first of these equations by $A$, dividing the second equation by $C$, and using Eq.81 gives,

$$\frac{\sin \theta_A}{A} = \frac{\sin \theta_C}{C} = \frac{h}{AC}. \tag{87}$$

Like with the acute triangle, showing that $\frac{\sin \theta_B}{B}$ is also equal to $\frac{h}{AC}$ will require a little more effort. We start by using the facts that the sum of the interior angles of any plane triangle

17

must equal 180 degrees[6] and that $\sin(180° - \theta) = \sin\theta$ to write,

$$
\begin{aligned}
\sin\theta_B &= \sin(180° - \theta_A - \theta_C) \\
&= \sin(\theta_A + \theta_C) \\
&= \sin\theta_A \cos\theta_C + \cos\theta_A \sin\theta_C \ ,
\end{aligned}
\tag{88}
$$

where, in the third line, we have used the first addition formula in Sec.2.2 (Eq.13). Using Eq.81 - Eq.86, the above equation reduces to,

$$
\begin{aligned}
\sin\theta_B &= -\frac{h}{C}\frac{B'}{A} + \frac{B+B'}{C}\frac{h}{A} \\
&= \frac{hB}{AC} \ .
\end{aligned}
\tag{89}
$$

Now, dividing this equation by $B$ gives,

$$
\frac{\sin\theta_B}{B} = \frac{h}{AC} \ ,
\tag{90}
$$

which proves the Law of Sines for obtuse triangles. We have now proven the Law of Sines for any triangle that can be drawn on a plane.

---

[6]See App.B for an explanation of why this is true.

## 2.5    The Law of Cosines

---

*The Law of Cosines*

For any angle triangle on a plane,

$$C^2 = A^2 + B^2 - 2AB\cos\theta_C \tag{91}$$

where $A$, $B$, and $C$ are the lengths of the sides opposite from the angles with corresponding subscripts (see Fig.7 and Fig.8). Similar relations hold for the other two angles.

---

Like the Law of Sines, the Law of Cosines is fairly obvious for right triangles. If $\theta_C$ is the right angle, $\cos\theta_C = 0$ and the Law of Cosines reduces to the Pythagorean Theorem (Eq.1). If $\theta_C$ is one of the other two angles, we can use the fact that $\cos\theta_C$ is equal to the length of the adjacent side divided by the hypotenuse and the Law of Cosines reduces to a rearranged version of the Pythagorean Theorem. To prove it holds for any plane triangle, we will, once again consider acute and obtuse triangles separately. For convenience, Fig.5 and Fig.6 have been reproduced below.

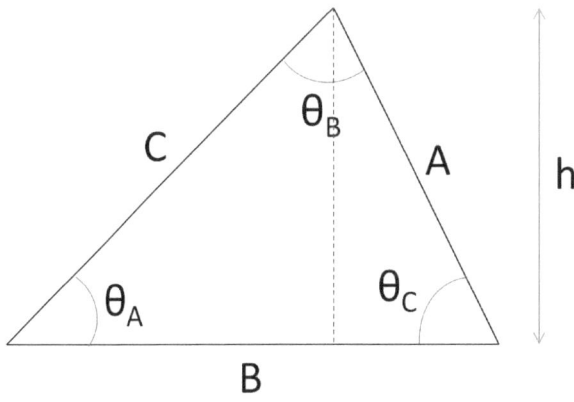

Figure 7: Acute triangle of height $h$.

Like before, we will split the side labeled $B$ into to parts with the part to the left of the dashed line having length $B_1$ and the part to the right of the dashed line having length $B_2$. From basic trigonometry, we know,

$$B_2 = A\cos\theta_C \,, \tag{92}$$

$$h = A\sin\theta_C \,. \tag{93}$$

Using these equations with the Pythagorean Theorem (Eq.1) we have,

$$
\begin{aligned}
C^2 &= B_1^2 + h^2 \\
&= (B - A\cos\theta_C)^2 + (A\sin\theta_C)^2 \\
&= B^2 - 2AB\cos\theta_C + A^2\cos^2\theta_C + A^2\sin^2\theta_C \\
&= A^2 + B^2 - 2AB\cos\theta_C \,,
\end{aligned}
\tag{94}
$$

where in the last line we have used the Fundamental Trigonometric Identity (Eq.6). The Law of Cosines can be proven for the other angles by simply rotating the triangle and performing the same procedure. This finishes the proof of the Law of Cosines for acute triangles.

To prove the Law of Cosines for obtuse triangles we will refer to Fig.8. First, we note again,

$$\sin\theta_{C'} = \sin\theta_C \,, \tag{95}$$

$$\cos\theta_{C'} = -\cos\theta_C \,. \tag{96}$$

Using these equations and basic trigonometry, we have,

$$h = A \sin \theta_{C'}$$

$$= A \sin \theta_C , \qquad (97)$$

$$B' = A \cos \theta_{C'}$$

$$= -A \cos \theta_C . \qquad (98)$$

Combining these equations with the Pythagorean Theorem (Eq.1) gives,

$$C^2 = h^2 + (B + B')^2$$

$$= (A \sin \theta_C)^2 + B^2 + 2B(-A \cos \theta_C) + (-A \cos \theta_C)^2$$

$$= A^2 \sin^2 \theta_C + B^2 - 2AB \cos \theta_C + A^2 \cos^2 \theta_C$$

$$= A^2 + B^2 - 2AB \cos \theta_C , \qquad (99)$$

where in the last line we have used the Fundamental Trigonometric Identity (Eq.6). This completes the proof of the Law of Cosines for the obtuse angle.

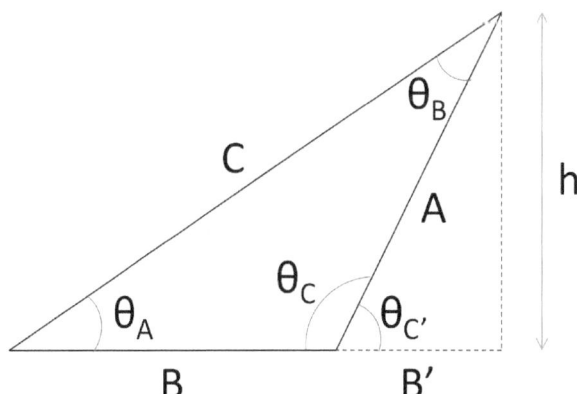

Figure 8: Obtuse triangle of height $h$.

To prove the Law of Cosines for the acute angle $\theta_A$, we focus on the right triangle formed by the dashed lines and the side labeled $A$. Using the Pythagorean Theorem (Eq.1) and basic trigonometry, we have,

$$
\begin{aligned}
A^2 &= B'^2 + h^2 \\
&= (C\cos\theta_A - B)^2 + (C\sin\theta_A)^2 \\
&= C^2\cos^2\theta_A + B^2 - 2BC\cos\theta_A + C^2\sin^2\theta_A \\
&= B^2 + C^2 - 2BC\cos\theta_A \ .
\end{aligned}
\tag{100}
$$

The Law of Cosines for $\theta_B$ can be proven in a similar way which completes the proof of the Law of Cosines for obtuse triangles. We have now proven the Law of Cosines for any triangle that can be drawn on a plane.

# 3 The Quadratic Formula

---

*The Quadratic Formula*

A quadratic equation of the form

$$Ax^2 + Bx + C = 0 \qquad (101)$$

has the solutions

$$x = \frac{-B \pm \sqrt{B^2 - 4AC}}{2A} \qquad (102)$$

---

To derive this formula, we will first simplify the problem by dividing Eq.101 by $A$ giving,

$$x^2 + \frac{B}{A}x + \frac{C}{A} = 0 \ . \qquad (103)$$

Any quadratic equation can be factored, even if the solutions are not integers, and can be written as,

$$(x - x_1)(x - x_2) = 0 \ , \qquad (104)$$

where $x_1$ and $x_2$ are the solutions. Expanding Eq.104 and matching the coefficients with Eq.103 gives,

$$x^2 - (x_1 + x_2)x + x_1 x_2 = 0 \ , \qquad (105)$$

with

$$x_1 + x_2 = -\frac{B}{A} \tag{106}$$

and

$$x_1 x_2 = \frac{C}{A} \, . \tag{107}$$

Attempting to solve this system of two equations by the usual method of substitution will result in a quadratic equation for which we will need a general solution. Since we are trying to derive a general solution, we need to find another method. We can do this by introducing a third equation with a third unknown. Let's call the difference between the solutions $D$,

$$x_1 - x_2 = D \, . \tag{108}$$

Solving Eq.106 for $x_2$ gives,

$$x_2 = -\frac{B}{A} - x_1 \, . \tag{109}$$

Substituting this into Eq.108 and simplifying gives,

$$2x_1 + \frac{B}{A} = D \, . \tag{110}$$

Using the above two equations, we now have,

$$x_1 = \frac{1}{2}\left(D - \frac{B}{A}\right) \tag{111}$$

and

$$x_2 = -\frac{1}{2}\left(D + \frac{B}{A}\right) . \tag{112}$$

We can now use Eq.107 to solve for $D$,

$$
\begin{aligned}
\left[\frac{1}{2}\left(D - \frac{B}{A}\right)\right]\left[-\frac{1}{2}\left(D + \frac{B}{A}\right)\right] &= \frac{C}{A} \\
-\frac{1}{4}\left(D^2 - \frac{B^2}{A^2}\right) &= \frac{C}{A} . \tag{113}
\end{aligned}
$$

Solving this equation for $D$ gives,

$$
\begin{aligned}
D &= \pm\sqrt{\frac{B^2}{A^2} - 4\frac{C}{A}} \\
&= \pm\frac{\sqrt{B^2 - 4AC}}{A} . \tag{114}
\end{aligned}
$$

Substituting this expression into Eq.111 and Eq.112 gives,

$$x_1 = \frac{1}{2}\left(\pm\frac{\sqrt{B^2 - 4AC}}{A} - \frac{B}{A}\right) \tag{115}$$

$$x_2 = -\frac{1}{2}\left(\pm\frac{\sqrt{B^2 - 4AC}}{A} + \frac{B}{A}\right) . \tag{116}$$

These solutions can written in the more compact form,

$$x = \frac{-B \pm \sqrt{B^2 - 4AC}}{2A} , \tag{117}$$

which is the quadratic formula.

## 3.1 Alternative Derivation

An easier but less intuitive way to derive the quadratic formula is to use the method of completing the square. Starting with Eq.103, we subtract the constant term from both sides of the equation giving,

$$x^2 + \frac{B}{A}x = -\frac{C}{A} \; . \tag{118}$$

Next, we divide the coefficient of the linear term by two and add its square to both sides of the equation to get,

$$x^2 + \frac{B}{A}x + \frac{B^2}{4A^2} = \frac{B^2}{4A^2} - \frac{C}{A} \; . \tag{119}$$

The right-hand side of the equation can now be written as a square and we have,

$$\left(x + \frac{B}{2A}\right)^2 = \frac{B^2}{4A^2} - \frac{C}{A} \; . \tag{120}$$

Taking the square-root and subtracting $\frac{B}{2A}$ gives the solution,

$$x = \pm\sqrt{\frac{B^2}{4A^2} - \frac{C}{A}} - \frac{B}{2A} \; , \tag{121}$$

or

$$x = \frac{-B \pm \sqrt{B^2 - 4AC}}{2A} \; , \tag{122}$$

which is the quadratic formula.

# 4 The Fundamental Theorem of Calculus

---

*The Fundamental Theorem of Calculus*

For any function defined on the interval $(a, b)$,

$$\int_a^b f(x)dx = F(b) - F(a) \tag{123}$$

where $F(a)$ is the antiderivative of $f(x)$ evaluated at $x = a$ and $F(b)$ is the antiderivative of $f(x)$ evaluated at $x = b$.

---

To prove this theorem, we start with the definitions of derivative and integral. A derivative is the rate-of-change of a function, $f(x)$, with respect to the variable, $x$, and is formally defined as,

$$\frac{d}{dx}f(x) \equiv \lim_{\Delta x \to 0} \frac{f(x + \Delta x) - f(x)}{\Delta x} . \tag{124}$$

A definite integral measures the area between a curve defined by the function, $f(x)$, and the $x$-axis on the interval between $x = a$ and $x = b$ and is formally defined as,

$$\int_a^b f(x)dx \equiv \lim_{N \to \infty} \frac{b - a}{N} \sum_{n=0}^N f\left(a + \frac{n(b - a)}{N}\right) . \tag{125}$$

$F(x)$ is the antiderivative of $f(x)$, which means,

$$f(x) \equiv \frac{d}{dx}F(x) . \tag{126}$$

Using the definition of derivative (Eq.124), we have,

$$f(x) = \lim_{\Delta x \to 0} \frac{F(x + \Delta x) - F(x)}{\Delta x} . \tag{127}$$

Substituting this expression into the right-hand-side of the definition of a definite integral (Eq.125) gives,

$$\int_a^b f(x)dx = \lim_{N \to \infty} \frac{b-a}{N} \sum_{n=0}^{N} \lim_{\Delta x \to 0} \frac{F\left(a + \frac{n(b-a)}{N} + \Delta x\right) - F\left(a + \frac{n(b-a)}{N}\right)}{\Delta x} . \tag{128}$$

With the identification,

$$\lim_{\Delta x \to 0} \Delta x = \lim_{N \to \infty} \frac{b-a}{N} , \tag{129}$$

the above equation reduces to,

$$\int_a^b f(x)dx = \lim_{N \to \infty} \sum_{n=0}^{N} \left[ F\left(a + \frac{(n+1)(b-a)}{N}\right) - F\left(a + \frac{n(b-a)}{N}\right) \right] . \tag{130}$$

This expression can be rewritten as two separate summations,

$$\int_a^b f(x)dx = \lim_{N \to \infty} \left[ \sum_{n=1}^{N+1} F\left(a + \frac{n(b-a)}{N}\right) - \sum_{n=0}^{N} F\left(a + \frac{n(b-a)}{N}\right) \right] , \tag{131}$$

where the summation index has been adjusted on the first term. The functions in each summation are identical, so the only terms that survive are the $n = N + 1$ term in the first summation and the $n = 0$ term in the second summation,

$$\int_a^b f(x)dx = \lim_{N \to \infty} \left[ F\left(a + \frac{(N+1)(b-a)}{N}\right) - F(a) \right] . \tag{132}$$

Now, taking the limit $N \to \infty$ gives,

$$\int_a^b f(x)dx = F(b) - F(a) , \qquad (133)$$

which is the Fundamental Theorem of Calculus.

Another, more intuitive but less rigorous way, to understand the Fundamental Theorem of Calculus is to simply substitute the right-hand-side of Eq.126 into the left hand side of Eq.123 and treat the differentials, $dx$, like finite numbers. This gives,

$$\begin{aligned}
\int_a^b f(x)dx &= \int_a^b \frac{d}{dx}F(x)dx \\
&= \int_a^b dF(x) .
\end{aligned} \qquad (134)$$

From the definition of integral (Eq.125), we know that the second line of the above equation is telling us to add up all the differentials of $F(x)$ between the points $x = a$ and $x = b$, which is just the difference between the function evaluated at these two points. This method is not a valid proof because there is no reason to think that differentials can be treated like finite numbers, however, in many cases this naive thinking leads to the correct answer.[7]

It is important to understand that the Fundamental Theorem of Calculus is far from obvious. Many students consider an integral to be just a way of denoting the antiderivative. While it's true that integrals are closely related to antiderivatives, this is not the definition. A derivative measures how quickly a function changes with respect to a variable and an integral is a summation of infinitely many infinitesimal terms. The fact that these very different concepts are so closely related is the power of theorem. It is called the Fundamental Theorem of Calculus because it forms the foundation for even more powerful theorems in calculus that we will now explore.

---

[7]This was a major point of contention between Isaac Newton and Gottfried Leibniz, who independently discovered calculus in the $17^{th}$ century.

# 5 The Gradient Theorem

---

*The Gradient Theorem*

For any differentiable function defined on the curve, $C$,

$$\int_C \vec{\nabla} f(x, y, z) \cdot d\vec{\ell} = f(x_f, y_f, z_f) - f(x_i, y_i, z_i) \qquad (135)$$

where the integral is evaluated along the curve, $C$, from the point, $(x, y, z) = (x_i, y_i, z_i)$, to the point, $(x, y, z) = (x_f, y_f, z_f)$, and $d\vec{\ell}$ is the differential distance measured along the curve.

---

This theorem can be proved using the same method we used to prove the Fundamental Theorem of Calculus. The biggest difference is the integral is evaluated along an arbitrary curve instead of the $x$-axis. In fact, The Fundamental Theorem of Calculus is a special case of the Gradient Theorem where $(x_i, y_i, z_i) = (a, 0, 0)$ and $(x_f, y_f, z_f) = (b, 0, 0)$. To make this more clear, we need to understand that the integrand, $\vec{\nabla} f(x, y, z) \cdot d\vec{\ell}$, is the differential of the function, $f(x, y, z)$, measured along the curve, $C$. The differential length, $d\vec{\ell}$, is defined as,

$$d\vec{\ell} \equiv d\vec{x} + d\vec{y} + d\vec{z} , \qquad (136)$$

and the gradient operator, $\vec{\nabla}$, is defined as,

$$\vec{\nabla} \equiv \frac{\partial}{\partial x} \hat{x} + \frac{\partial}{\partial y} \hat{y} + \frac{\partial}{\partial z} \hat{z} , \qquad (137)$$

where $\hat{x}$, $\hat{y}$, and $\hat{z}$ are unit vectors representing the $x$, $y$, and $z$, directions respectively.

Remember, unit vectors in rectangular coordinates[8] are constant so they are not affected by the derivatives but this is not the case in all coordinate systems (for example, unit vectors in spherical coordinates are not constant).

Applying the dot product gives,

$$\vec{\nabla} f(x,y,z) \cdot d\vec{\ell} = \frac{\partial}{\partial x} f(x,y,z) dx + \frac{\partial}{\partial y} f(x,y,z) dy + \frac{\partial}{\partial z} f(x,y,z) dz . \tag{138}$$

The first term in the above expression is the change in $f(x,y,z)$ due to the change in $x$, the second term is the change in $f(x,y,z)$ due to the change in $y$, and the third term is the change in $f(x,y,z)$, due to the change in $z$. Adding these three terms together gives the total differential change in the function, $f(x,y,z)$, so we can write,

$$\vec{\nabla} f(x,y,z) \cdot d\vec{\ell} = df . \tag{139}$$

The validity of the Gradient Theorem now follows from the definition of integral (see the simplified explanation of the Fundamental Theorem of Calculus at the end of the previous section).

The Gradient Theorem can easily be generalized to an arbitrary number of variables, as long as the integral is evaluated along a well-defined curve. In one dimension, the Gradient Theorem reduces to the Fundamental Theorem of Calculus. We can now use the Gradient Theorem to prove theorems involving multidimensional integrals.

---

[8]Rectangular coordinates are also called Cartesian coordinates after the $17^{th}$ century mathematician, René Descartes.

# 6 The Curl Theorem

*The Curl Theorem*[9]

For a vector function, $\vec{F}$, defined on the surface, $S$, bounded by the curve, $C$,

$$\int_S \vec{\nabla} \times \vec{F} \cdot d\vec{\alpha} = \oint_C \vec{F} \cdot d\vec{\ell} \tag{140}$$

where $d\vec{\alpha}$ is a differential piece of the surface and $d\vec{\ell}$ is a differential piece of the bounding curve. The circle in the middle of the integral sign on the right side indicates that the curve, $C$, is closed. The integral is evaluated according to a right-hand-rule. If $d\vec{\alpha}$ points out of the page, the line integral is evaluated in the counter-clockwise direction.

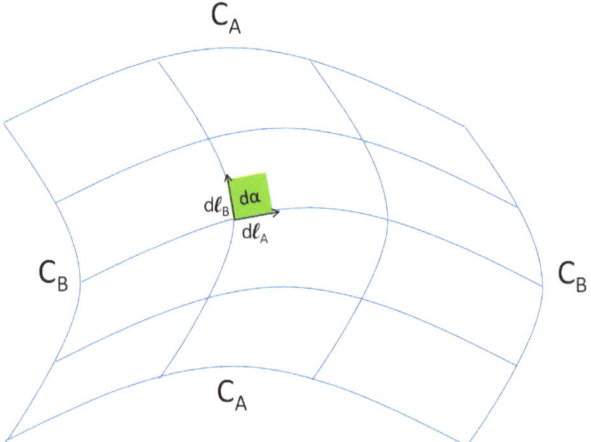

Figure 9: An arbitrary surface is formed by the intersections of two sets of curves labeled $C_A$ and $C_B$. A differential area, $d\vec{\alpha}$, is formed by the cross-product of two differential lengths, $d\vec{\ell}_A$ and $d\vec{\ell}_B$, where $d\vec{\ell}_A$ is along one of curves labeled $C_A$ and $d\vec{\ell}_B$ is along one of the curves labeled $C_B$.

---

[9]The Curl Theorem is sometimes referred to as Stokes' Theorem.

To prove this theorem, we will need to refer to Fig.9. The upper-most, lower-most, and all similarly shaped curves between them are labeled $C_A$. The left-most, right-most, and all similarly shaped curves between them are labeled $C_B$. The surface, $S$, is formed by the intersections of these two sets of curves.

A differential area, $d\vec{\alpha}$, is formed by the cross-product of differential lengths along intersecting curves,

$$d\vec{\alpha} = d\vec{\ell}_A \times d\vec{\ell}_B .$$
(141)

The cross-product is defined as,

$$d\vec{\ell}_A \times d\vec{\ell}_B \equiv (dy_A dz_B - dz_A dy_B)\,\hat{x} + (dz_A dx_B - dx_A dz_B)\,\hat{y} + (dx_A dy_B - dy_A dx_B)\,\hat{z} ,$$
(142)

where $\hat{x}$, $\hat{y}$, and $\hat{z}$ represent unit vectors. The differential area is a vector perpendicular to the surface, $S$. The curl of a vector field is defined as,

$$\vec{\nabla} \times \vec{F} \equiv \left(\frac{\partial F_z}{\partial y} - \frac{\partial F_y}{\partial z}\right)\hat{x} + \left(\frac{\partial F_x}{\partial z} - \frac{\partial F_z}{\partial x}\right)\hat{y} + \left(\frac{\partial F_y}{\partial x} - \frac{\partial F_x}{\partial y}\right)\hat{z} .$$
(143)

Substituting the above three equations into the left side of Eq.140 and performing the dot-product gives,

$$
\begin{aligned}
\int_S \vec{\nabla} \times \vec{F} \cdot d\vec{\alpha} &= \int_S \Bigg[ \left(\frac{\partial F_z}{\partial y} - \frac{\partial F_y}{\partial z}\right)\hat{x} + \left(\frac{\partial F_x}{\partial z} - \frac{\partial F_z}{\partial x}\right)\hat{y} + \left(\frac{\partial F_y}{\partial x} - \frac{\partial F_x}{\partial y}\right)\hat{z} \Bigg] \\
&\quad \cdot \Big[ (dy_A dz_B - dz_A dy_B)\,\hat{x} + (dz_A dx_B - dx_A dz_B)\,\hat{y} + (dx_A dy_B - dy_A dx_B)\,\hat{z} \Big] \\
&= \int_S \Bigg[ \left(\frac{\partial F_z}{\partial y} - \frac{\partial F_y}{\partial z}\right)(dy_A dz_B - dz_A dy_B) \\
&\quad + \left(\frac{\partial F_x}{\partial z} - \frac{\partial F_z}{\partial x}\right)(dz_A dx_B - dx_A dz_B) \\
&\quad + \left(\frac{\partial F_y}{\partial x} - \frac{\partial F_x}{\partial y}\right)(dx_A dy_B - dy_A dx_B) \Bigg] .
\end{aligned}
$$
(144)

We now apply the distributive property to each set of parentheses to get,

$$
\int_S \vec{\nabla} \times \vec{F} \cdot d\vec{\alpha} = \int_S \left[ \frac{\partial F_z}{\partial y} dy_A dz_B - \frac{\partial F_z}{\partial y} dz_A dy_B - \frac{\partial F_y}{\partial z} dy_A dz_B + \frac{\partial F_y}{\partial z} dz_A dy_B \right.
$$
$$
+ \frac{\partial F_x}{\partial z} dz_A dx_B - \frac{\partial F_x}{\partial z} dx_A dz_B - \frac{\partial F_z}{\partial x} dz_A dx_B + \frac{\partial F_z}{\partial x} dx_A dz_B
$$
$$
\left. + \frac{\partial F_y}{\partial x} dx_A dy_B - \frac{\partial F_y}{\partial x} dy_A dx_B - \frac{\partial F_x}{\partial y} dx_A dy_B + \frac{\partial F_x}{\partial y} dy_A dx_B \right] . \quad (145)
$$

Adding $\frac{\partial F_x}{\partial x} dx_A dx_B + \frac{\partial F_y}{\partial y} dy_A dy_B + \frac{\partial F_z}{\partial z} dz_A dz_B - \frac{\partial F_x}{\partial x} dx_A dx_B - \frac{\partial F_y}{\partial y} dy_A dy_B - \frac{\partial F_z}{\partial z} dz_A dz_B$ to the right side of the above equation, then factoring $dx_B$, $dy_B$, or $dz_B$ out of the positive terms and factoring $dx_A$, $dy_A$, or $dz_A$ out of the negative terms, we have,

$$
\int_S \vec{\nabla} \times \vec{F} \cdot d\vec{\alpha} = \int_S \left[ \left( \frac{\partial F_x}{\partial x} dx_A + \frac{\partial F_x}{\partial y} dy_A + \frac{\partial F_x}{\partial z} dz_A \right) dx_B \right.
$$
$$
+ \left( \frac{\partial F_y}{\partial x} dx_A + \frac{\partial F_y}{\partial y} dy_A + \frac{\partial F_y}{\partial z} dz_A \right) dy_B
$$
$$
+ \left( \frac{\partial F_z}{\partial x} dx_A + \frac{\partial F_z}{\partial y} dy_A + \frac{\partial F_z}{\partial z} dz_A \right) dz_B
$$
$$
- \left( \frac{\partial F_x}{\partial x} dx_B + \frac{\partial F_x}{\partial y} dy_B + \frac{\partial F_x}{\partial z} dz_B \right) dx_A
$$
$$
- \left( \frac{\partial F_y}{\partial x} dx_B + \frac{\partial F_y}{\partial y} dy_B + \frac{\partial F_y}{\partial z} dz_B \right) dy_A
$$
$$
\left. - \left( \frac{\partial F_z}{\partial x} dx_B + \frac{\partial F_z}{\partial y} dy_B + \frac{\partial F_z}{\partial z} dz_B \right) dz_A \right] . \quad (146)
$$

The left side of the equation is unchanged because the sum of the added terms is zero. This equation can now be written in a more compact way using gradient operators and dot-products,

$$
\int_S \vec{\nabla} \times \vec{F} \cdot d\vec{\alpha} = \int_S \left[ \vec{\nabla} \left( \vec{F} \cdot d\vec{\ell}_B \right) \cdot \vec{\ell}_A - \vec{\nabla} \left( \vec{F} \cdot d\vec{\ell}_A \right) \cdot \vec{\ell}_B \right] . \quad (147)
$$

Since integrating over the surface, $S$, is the same as integrating over both sets of curves, $C_A$ and $C_B$, we can replace $\int_S$ with $\int_{C_A} \int_{C_B}$ or $\int_{C_B} \int_{C_A}$,

$$\int_S \vec{\nabla} \times \vec{F} \cdot d\vec{\alpha} = \int_{C_A} \int_{C_B} \vec{\nabla} \left( \vec{F} \cdot d\vec{\ell}_B \right) \cdot \vec{\ell}_A - \int_{C_B} \int_{C_A} \vec{\nabla} \left( \vec{F} \cdot d\vec{\ell}_A \right) \cdot \vec{\ell}_B . \quad (148)$$

The order of integration and differentiation operations is irrelevant,[10] so we can move the gradient operators outside the inner integrals,

$$\int_S \vec{\nabla} \times \vec{F} \cdot d\vec{\alpha} = \int_{C_A} \vec{\nabla} \left( \int_{C_B} \vec{F} \cdot d\vec{\ell}_B \right) \cdot \vec{\ell}_A - \int_{C_B} \vec{\nabla} \left( \int_{C_A} \vec{F} \cdot d\vec{\ell}_A \right) \cdot \vec{\ell}_B . \quad (149)$$

Using the Gradient Theorem (Eq.135) to evaluate the outer integrals gives,

$$\int_S \vec{\nabla} \times \vec{F} \cdot d\vec{\alpha} = \left( \int_{C_B} \vec{F} \cdot d\vec{\ell}_B \right) \Big|_{C_{A_i}}^{C_{A_f}} - \left( \int_{C_A} \vec{F} \cdot d\vec{\ell}_A \right) \Big|_{C_{B_i}}^{C_{B_f}} , \quad (150)$$

where the first integral is evaluated at the initial and final points on the curves labeled $C_A$ and the second integral is evaluated at the initial and final points of the curves labeled $C_B$. The initial points on the curves labeled $C_A$ are on the left-most curve labeled $C_B$, the final points on the curves labeled $C_A$ are on the right-most curve labeled $C_B$, the initial points on the curves labeled $C_B$ are on the lower-most curve labeled $C_A$, and the final points on the curves labeled $C_B$ are on the upper-most curve labeled $C_A$. We can now rewrite the above equation as,

$$\int_S \vec{\nabla} \times \vec{F} \cdot d\vec{\alpha} = \int_{C_{B_R}} \vec{F} \cdot d\vec{\ell}_B - \int_{C_{B_L}} \vec{F} \cdot d\vec{\ell}_B - \int_{C_{A_U}} \vec{F} \cdot d\vec{\ell}_A + \int_{C_{A_L}} \vec{F} \cdot d\vec{\ell}_A , \quad (151)$$

where $C_{B_L}$ is the left-most curve labeled $C_B$, $C_{B_R}$ is the right-most curve labeled $C_B$, $C_{A_L}$ is the lower-most curve labeled $C_A$, and $C_{A_U}$ is the upper-most curve labeled $C_A$. Combined, these four curves make the curve, $C$, the perimeter of the surface, $S$. The negative signs in

---

[10]This is not always true. To see that it works in this case, just consider the definitions in Eq.124 and Eq.125.

front of the second and third terms can be changed to plus signs if the integrals are evaluated in the opposite direction. Doing this and rearranging the terms gives,

$$\int_S \vec{\nabla} \times \vec{F} \cdot d\vec{\alpha} = \int_{C_{B_R}} \vec{F} \cdot d\vec{\ell}_B + \int_{C_{B_L}} \vec{F} \cdot \left( -d\vec{\ell}_B \right) + \int_{C_{A_U}} \vec{F} \cdot \left( -d\vec{\ell}_A \right) + \int_{C_{A_L}} \vec{F} \cdot d\vec{\ell}_A$$

$$= \int_{C_{B_R}} \vec{F} \cdot d\vec{\ell}_B + \int_{C_{A_U}} \vec{F} \cdot \left( -d\vec{\ell}_A \right) + \int_{C_{B_L}} \vec{F} \cdot \left( -d\vec{\ell}_B \right) + \int_{C_{A_L}} \vec{F} \cdot d\vec{\ell}_A \,. \quad (152)$$

These four line integrals can now be evaluated continuously around the surface, $S$, along the curve, $C$, and we have,

$$\int_S \vec{\nabla} \times \vec{F} \cdot d\vec{\alpha} = \oint_C \vec{F} \cdot d\vec{\ell} \,, \quad (153)$$

which is the Curl Theorem.

# 7  The Divergence Theorem

---

*The Divergence Theorem*[11]

For a vector function, $\vec{F}$, defined in the volume, $V$, bounded by the surface, $S$,

$$\int_V \vec{\nabla} \cdot \vec{F} \, dV = \oint_S \vec{F} \cdot d\vec{\alpha} \tag{154}$$

where $dV$ is a differential piece of the volume and $d\vec{\alpha}$ is a differential piece of the bounding surface. The vector, $d\vec{\alpha}$, points away from the interior of the volume.

---

To prove this theorem, we first separate the left side of Eq.154 into three parts,

$$
\begin{aligned}
\int_V \vec{\nabla} \cdot \vec{F} \, dV &= \int_V \left( \frac{\partial F_x}{\partial x} + \frac{\partial F_y}{\partial y} + \frac{\partial F_z}{\partial z} \right) dx dy dz \\
&= \int_V \frac{\partial F_x}{\partial x} \, dx dy dz + \int_V \frac{\partial F_y}{\partial y} \, dx dy dz + \int_V \frac{\partial F_z}{\partial z} \, dx dy dz \; .
\end{aligned}
\tag{155}
$$

Focusing on the first term and integrating over $x$ gives,

$$\int_V \frac{\partial F_x}{\partial x} \, dx dy dz = \int \int F_x \, dy dz \; . \tag{156}$$

The differential area, $dy dz$, is the magnitude of the $x$-component of $d\vec{\alpha}$ so we may write,

$$\int_V \frac{\partial F_x}{\partial x} \, dx dy dz = \int_S F_x \, d\alpha_x \; . \tag{157}$$

---

[11]The Divergence Theorem is sometimes referred to as Gauss's Theorem or Green's Theorem.

Similarly,

$$\int_V \frac{\partial F_y}{\partial y}\, dxdydz = \int_S F_y\, d\alpha_y \,, \tag{158}$$

$$\int_V \frac{\partial F_z}{\partial z}\, dxdydz = \int_S F_z\, d\alpha_z \,. \tag{159}$$

Substituting the above three equations into Eq.155 gives,

$$\int_V \vec{\nabla} \cdot \vec{F}\, dV = \int F_x\, d\alpha_x + \int F_y\, d\alpha_y + \int F_z\, d\alpha_z \,, \tag{160}$$

or,

$$\int_V \vec{\nabla} \cdot \vec{F}\, dV = \oint_S F \cdot d\vec{\alpha} \,, \tag{161}$$

which is the Divergence Theorem.

# 8    Taylor's Theorem

*Taylor's Theorem*

Any differentiable function, $f(x)$, can be expanded about $x = a$ according to the formula,

$$f(x) \; = \; \sum_{n=0}^{\infty} \frac{(x-a)^n}{n!} f^{(n)}(a)$$

$$= \; f(a) + (x-a) f'(a) + \frac{(x-a)^2}{2!} f''(a) + \ldots \tag{162}$$

where $f^{(n)}(a)$ is the $n^{th}$ derivative of $f(x)$ evaluated at $x = a$.[12]

Although this is a very powerful theorem, its validity becomes clear if one considers the most simplistic way to estimate a function near a particular value. The most naive way to estimate the value of a function, $f(x)$, near $x = a$ is to assume the function does not change much and write $f(x) \approx f(a)$. Let's call this the zeroth-order approximation. Obviously, this approximation will be very bad for most functions unless $x \approx a$, so it is useful to consider a correction, $\Delta f(x) = f(x) - f(a)$. From the Fundamental Theorem of Calculus we know,

$$\Delta f(x) \; = \; \int_{a}^{x} \frac{df}{dx} \, dx$$

$$= \; \int_{a}^{x} f'(x) \, dx \; . \tag{163}$$

To get the first-order approximation, we need an approximation for $f'(x)$. Let's use the same naive approach we used to get the zeroth-order approximation to $f(x)$ and write $f'(x) \approx f'(a)$. Of course, to improve the approximation, we need to include another correction.

---

[12]This formula is often referred to as a Taylor series or a power series. If $a = 0$ it is often referred to as a Maclaurin series. There may be some convergence issues if the series is evaluated at a value of $x$ that is not close enough to $a$ but we will not deal with those in this book.

Again, we can use the Fundamental Theorem of Calculus.

$$\begin{aligned}
\Delta f'(x) &= f'(x) - f'(a) \\
&= \int_a^x f''(x)\, dx \ .
\end{aligned}$$
(164)

We now have,

$$\begin{aligned}
\Delta f(x) &= \int_a^x \left( f'(a) + \int_a^x f''(x) dx \right) dx \\
&= (x - a) f'(a) + \int_a^x \left( \int_a^x f''(x) dx \right) dx \ .
\end{aligned}$$
(165)

Continuing in this way gives,

$$\begin{aligned}
\int_a^x \left( \int_a^x f''(x) dx \right) dx &= \int_a^x \left( \int_a^x \left( f''(a) + \int_a^x f'''(x) dx \right) dx \right) dx \\
&= \int_a^x \left( (x - a) f''(a) + \int_a^x \left( \int_a^x f'''(x) dx \right) dx \right) dx \\
&= \frac{1}{2}(x - a)^2 f''(a) + \int_a^x \left( \int_a^x \left( \int_a^x f'''(x) dx \right) dx \right) dx \ ,
\end{aligned}$$
(166)

$$\begin{aligned}
\int_a^x \left( \int_a^x \left( \int_a^x f'''(x) dx \right) dx \right) dx &= \int_a^x \left( \frac{1}{2}(x - a)^2 f'''(a) + \ldots \right) dx \\
&= \frac{(x - a)^3}{2 \cdot 3} f'''(a) + \ldots \ .
\end{aligned}$$
(167)

The next term in the series will be $\frac{(x-a)^4}{4 \cdot 3 \cdot 2} f^{(4)}(a)$. From this, one can clearly see the pattern emerging. Putting all the pieces together we have Taylor's Theorem,

$$\begin{aligned}
f(x) &= f(a) + (x - a) f'(a) + \frac{(x - a)^2}{2} f''(a) + \frac{(x - a)^3}{2 \cdot 3} f'''(a) + \frac{(x - a)^4}{4 \cdot 3 \cdot 2} f^{(4)}(a) + \ldots \\
&= \sum_{n=0}^{\infty} \frac{(x - a)^n}{n!} f^{(n)}(a) \ .
\end{aligned}$$
(168)

# 9 Euler's Formula

$$e^{i\theta} = \cos\theta + i\sin\theta \qquad (169)$$

This theorem can easily be proved with Taylor's Theorem. Expanding $e^{i\theta}$ in a Taylor series about $a = 0$ gives,

$$e^{i\theta} = \sum_{n=0}^{\infty} \frac{(i\theta)^n}{n!} \ . \qquad (170)$$

This sum can be separated into even and odd terms.

$$e^{i\theta} = \sum_{n=0}^{\infty} \frac{(i\theta)^{2n}}{(2n)!} + \sum_{n=0}^{\infty} \frac{(i\theta)^{2n+1}}{(2n+1)!} \ . \qquad (171)$$

Using $i \equiv \sqrt{-1}$, we have,

$$i^{2n} = (-1)^n \ , \qquad (172)$$

and

$$i^{2n+1} = i(-1)^n \ . \qquad (173)$$

Now the summations can be written,

$$e^{i\theta} = \sum_{n=0}^{\infty} (-1)^n \frac{(\theta)^{2n}}{(2n)!} + i \sum_{n=0}^{\infty} (-1)^n \frac{(\theta)^{2n+1}}{(2n+1)!} \ . \qquad (174)$$

Identifying the first summation as the Taylor series expansion of $\cos\theta$ and the second summation as the Taylor series expansion of $\sin\theta$ verifies Euler's Formula.

One thing that makes Euler's Formula useful is we can now express trigonometric functions in terms of complex exponentials. To show this, we first consider the complex conjugate,

$$e^{-i\theta} = \cos\theta - i\sin\theta \, , \tag{175}$$

where we have used the fact that $\cos\theta$ is an even function $(\cos(-\theta) = \cos\theta)$ and $\sin\theta$ is an odd function $(\sin(-\theta) = -\sin\theta)$. Taking the sum and difference of $e^{i\theta}$ and $e^{-i\theta}$ gives,

$$e^{i\theta} + e^{-i\theta} = 2\cos\theta \, , \tag{176}$$

and

$$e^{i\theta} - e^{-i\theta} = 2i\sin\theta \, . \tag{177}$$

These equations can be rewritten as,

$$\cos\theta = \frac{e^{i\theta} + e^{-i\theta}}{2} \, , \tag{178}$$

and

$$\sin\theta = \frac{e^{i\theta} - e^{-i\theta}}{2i} \, . \tag{179}$$

# 10    The Rationality of Square-Roots

*The Rationality of Square-Roots*

If $x$ is a positive integer, the square-root of $x$ must either be an integer or irrational.[13]

We start by focusing on prime numbers. It has been known since the days of Pythagoras (and maybe even earlier) that $\sqrt{2}$ is irrational.[14] We will prove that the square-root of any prime number is irrational by assuming the opposite and finding a contradiction. If $x$ is a rational number we can write,

$$\sqrt{x} = \frac{a}{b},\qquad(180)$$

where $a$ and $b$ are integers. We also assume that the fraction $\frac{a}{b}$ is in lowest terms. This means that if $\sqrt{x} = \frac{m}{n}$, where $m$ and $n$ are positive integers, it must be true that $a \leq m$ and $b \leq n$. Squaring both sides of Eq.180 and rearranging gives,

$$b^2 = \frac{a^2}{x}.\qquad(181)$$

Since $b$ is an integer, $b^2$ must also be an integer so $a^2$ is divisible by $x$. Since $x$ is prime, $a$ must also be divisible by $x$, which means we can write,

$$a = xm,\qquad(182)$$

---

[13] A rational number is one that can be expressed as a ratio of integers.
[14] This apparently caused distress for mathematicians in those days because it implies that an isosceles right triangle cannot be formed by placing sticks of equal length end to end.

where $m$ is a positive integer. Substituting this expression into the previous equation gives,

$$
\begin{aligned}
b^2 &= \frac{x^2 m^2}{x} \\
&= x m^2 \ ,
\end{aligned}
\tag{183}
$$

or

$$
\frac{b^2}{x} = m^2 \ .
\tag{184}
$$

Since $m$ is an integer, $m^2$ must also be an integer so $b^2$ is divisible by $x$. Since $x$ is prime, $b$ must also be divisible by $x$, which means we can write,

$$
b = xn \ ,
\tag{185}
$$

where $n$ is a positive integer. Substituting Eq.182 and Eq.185 into Eq.180 gives

$$
\begin{aligned}
\sqrt{x} &= \frac{xm}{xn} \\
&= \frac{m}{n} \ ,
\end{aligned}
\tag{186}
$$

where $m < a$ and $n < b$ unless $x = 1$ which contradicts our original assumption that the fraction $\frac{a}{b}$ is in lowest terms, therefore, $\sqrt{x}$ must be irrational if $x$ is prime.[15]

We will now extend the discussion to include composite numbers (numbers that are not prime). We will need to use prime factorization which says that any composite number can be expressed as a product of prime numbers.[16] We start with the special case where the

---

[15]Obviously, $\sqrt{1}$ is not irrational but this is just one of many reasons why mathematicians do not consider 1 to be a prime number even though it appears to satisfy the definition.

[16]Although the importance of prime factorization in number theory cannot be overstated, the fact that it exists is fairly obvious. By definition, a composite number can be factored into at least two other numbers. If those factors are not prime, they can also be factored. One can proceed in this way until all the factors are prime. The real power of prime factorization is the fact that it is unique which is not so obvious but that is not important for our purposes so it will not be proved here.

prime factorization of $x$ does not contain any exponents greater than 1. Assume,

$$x = p_1 p_2 \ldots , \tag{187}$$

where all the $p_i$ are different prime numbers. Squaring both sides of Eq.180, rearranging and substituting the above expression gives,

$$
\begin{aligned}
b^2 &= \frac{a^2}{x} \\
&= \frac{a^2}{p_1 p_2 \ldots} .
\end{aligned}
\tag{188}
$$

Since $b^2$ is an integer and all the $p_i$ are prime, $a$ must be divisible by all $p_i$. If $a$ is divisible by all $p_i$, it must also be divisible by $x$. From here, the discussion proceeds exactly as above and we find that we can write,

$$a = mx \tag{189}$$

$$b = nx , \tag{190}$$

where $m$ and $n$ are positive integers. Substituting these expressions into Eq.180 again leads to a contradiction of our assumption that the fraction $\frac{a}{b}$ is in lowest terms and we conclude that $\sqrt{x}$ is irrational if the prime factorization of $x$ contains no exponents greater than 1.

We now generalize to the case where the prime factorization may contain exponents greater than 1. Assume,

$$x = p_1^{n_1} p_2^{n_2} \ldots \tag{191}$$

where all the $p_i$ are prime numbers and all the $n_i$ are positive integers. If all the exponents are even numbers, $x$ is a perfect square and $\sqrt{x}$ is an integer. If the prime factorization contains odd exponents, we can factor $x$ into two numbers, $y$ and $z$, such that the prime factorization

45

of $y$ contains no exponent greater than 1 and the prime factorization of $z$ contains only even exponents. Assume,

$$x \ = \ yz \ , \tag{192}$$

with

$$y \ = \ p_1 p_2 ... \tag{193}$$

$$z \ = \ q_1^{m_1} q_2^{m_2} ... \ , \tag{194}$$

where all the $p_i$ are different prime numbers, all the $q_i$ are prime numbers, and all the $m_i$ are positive even integers. Taking the square-root of $x$ gives,

$$\sqrt{x} = \sqrt{y}\sqrt{z} \ . \tag{195}$$

Since the prime factorization of $y$ contains no exponents greater than 1 and the prime factorization of $z$ contains only even exponents, we know that $y$ must be irrational and $z$ must be an integer. The product of an irrational number and an integer (or any rational number) is irrational so we conclude that $\sqrt{x}$ must either be irrational or an integer (when $x$ is a perfect square) for any integer value of $x$.

# A    Small Angle Approximations

---

*Small Angle Approximations*

The following approximations are valid for small values of $\phi$:

$$\sin \phi \approx \phi \tag{196}$$

$$\cos \phi \approx 1 - \frac{1}{2}\phi^2 \tag{197}$$

---

The above approximations can be easily verified using Taylor's Theorem, however, doing so would require taking derivatives of trigonometric functions and the easiest derivation of these derivatives requires the use of the small angle approximations so it is useful to find another way to prove the validity of these approximations. The angle, $\phi$, measured with respect to the positive $y$-axis in the clockwise direction,[17] is defined with the unit circle shown in Fig.10. With this definition, we can write,

$$x = \sin \phi , \tag{198}$$

$$y = \cos \phi . \tag{199}$$

Using the equation for a circle (Eq.7) and solving for $y$ in the upper half-plane, we can also write,

$$y = \sqrt{1 - x^2} . \tag{200}$$

---

[17]We could have defined the angle in the same way as Fig.2 but this way allows for small angles to correspond to small values of $x$ and the Taylor series is expanded about $x = 0$ instead of $x = 1$.

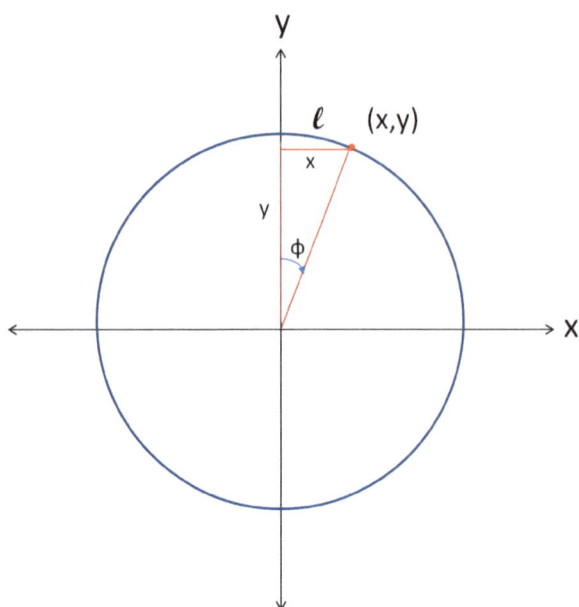

Figure 10: A unit circle centered at the origin in the $xy$ plane.

Since we are only concerned with small values of $\phi$, and therefore small values of $x$, we can expand this equation in a Taylor series about $x = 0$ giving,

$$y \approx 1 - \frac{1}{2}x^2 \ . \tag{201}$$

Next, we need to calculate the arc-length, $\ell$, defined by the angle, $\phi$. To do this, we will integrate the differential length, $d\ell$, along the circle, through the angle, $\phi$. Using the Pythagorean Theorem, we can express $d\ell$ in terms of the differentials, $dx$ and $dy$, where $dy$

can be expressed in terms of $dx$ using Eq.200. We now have,

$$
\begin{aligned}
\ell &= \int d\ell \\
&= \int \sqrt{dx^2 + dy^2} \\
&= \int \sqrt{dx^2 + \left(\frac{-x}{\sqrt{1-x^2}}\right)^2 dx^2} \\
&= \int \sqrt{1 + \frac{x^2}{1-x^2}}\, dx \\
&= \int \frac{1}{\sqrt{1-x^2}}\, dx \; . 
\end{aligned}
\tag{202}
$$

To simplify the integration, we can use the fact that we are only integrating over small values of $x$ and expand the integrand in a Taylor series. The result is,

$$
\begin{aligned}
\ell &\approx \int \left(1 + 3x^2\right) dx \\
&\approx x + x^3 \; .
\end{aligned}
\tag{203}
$$

Remembering, once again, that we are dealing with small values of $x$, we can ignore the second term and conclude,

$$
\ell \approx x \; .
\tag{204}
$$

It may seem strange to do all this work to approximate the arc-length when an exact expression is very easy to obtain. If $\phi$ is measured in radians, we have,[18]

$$
\ell = \phi \; .
\tag{205}
$$

---

[18]In general, the expression for arc-length is $\ell = r\phi$, where $r$ is the radius of the circle.

The reason for the approximation is that now we combine the above two expressions to obtain,

$$x \approx \phi , \tag{206}$$

for small values of $x$ and $\phi$. Combining this expression with Eq.198, Eq.199, and Eq.201 gives the small-angle approximations,

$$\sin \phi \approx \phi , \tag{207}$$

$$\cos \phi \approx 1 - \frac{1}{2}\phi^2 . \tag{208}$$

# B  Geometry

We start by considering Fig.11. Although we will not prove it here, it should not be difficult to convince yourself that the angles labeled $\theta$ and $\phi$ are equal if the horizontal lines are parallel. Just imagine moving the parallel lines closer together. When the lines overlap, the two angles must be the same.

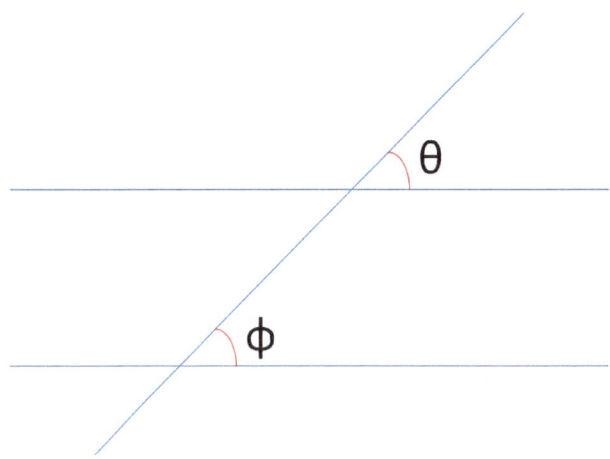

Figure 11: The two horizontal lines are parallel and intersected by a third line. The angles $\theta$ and $\phi$ are equal.

We will now show that opposite angles formed by intersecting lines are equal. Consider Fig.12. Combined, the angles labeled $\alpha$ and $\theta$ form a straight line, so by definition, their sum must equal 180 degrees,

$$\alpha + \theta = 180° \ . \tag{209}$$

Similarly,

$$\alpha + \phi = 180° \ , \tag{210}$$

since they also form a straight line. Solving Eq.209 for $\alpha$ and substituting into Eq.210 gives,

$$180° - \theta + \phi = 180° . \tag{211}$$

Subtracting 180° from both sides, we have,

$$-\theta + \phi = 0 , \tag{212}$$

or,

$$\theta = \phi . \tag{213}$$

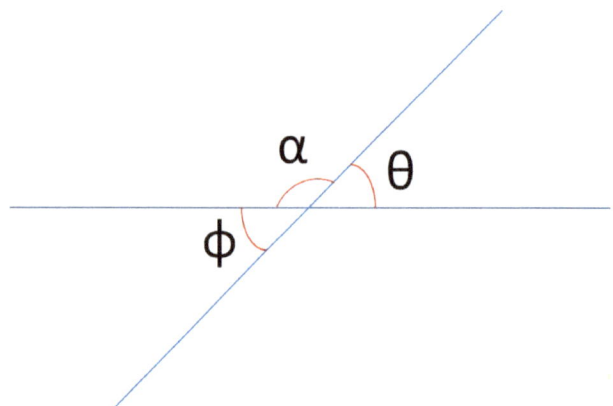

Figure 12: The angles $\theta$ and $\phi$ are equal.

We will now use these facts to show that the sum of the interior angles of a plane triangle is 180 degrees. Consider the triangle shown in Fig.13. The two angles labeled $\theta_1$ must be equal because they are formed by the intersections of one straight line with two parallel lines, as in Fig.11. Similarly, for the angles labeled $\theta_3$. The angles labeled $\theta_2$ must be equal because they are opposite angles formed by intersecting lines, as in Fig.12. Since the three angles form a straight line, their sum must be 180 degrees.

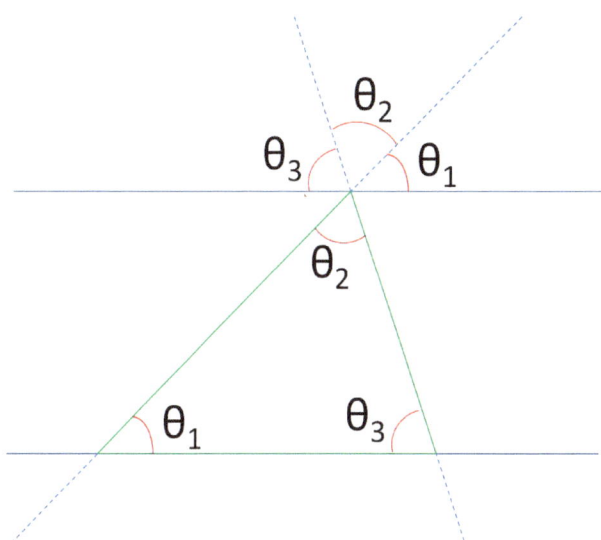

Figure 13: The interior angles can be rearranged to form a straight line showing their sum must be 180 degrees.

We can extend this reasoning to other figures by noting that any polygon can be formed by piecing together triangles. For example, any quadrilateral can be broken into two triangles by drawing a line through any two non-adjacent vertices. Any pentagon can be broken into a quadrilateral and a triangle by drawing a line through any two vertices with exactly one vertex between them. The minimum number of triangles needed to construct an $N$-sided figure is equal to $N - 2$. Each triangle contributes 180 degrees to the sum of interior angles so we conclude that the sum of the interior angles of any $N$-sided figure drawn on a plane must be $(N - 2)\,180°$.

# C    Trigonometric Derivatives

$$\frac{d}{dx}\sin x = \cos x \tag{214}$$

$$\frac{d}{dx}\cos x = -\sin x \tag{215}$$

To derive these expressions we will need to use the definition of derivative, trigonometric difference formulas, and the small angle approximations. From the definition of derivative (Eq.124) we have,

$$\frac{d}{dx}\sin x = \lim_{\Delta x \to 0}\frac{\sin(x+\Delta x)-\sin x}{\Delta x}\,, \tag{216}$$

and

$$\frac{d}{dx}\cos x = \lim_{\Delta x \to 0}\frac{\cos(x+\Delta x)-\cos x}{\Delta x}\,. \tag{217}$$

Using the difference formulas (Eq.39 and Eq.41) gives,

$$\frac{d}{dx}\sin x = \lim_{\Delta x \to 0}\frac{2\sin\left(\frac{\Delta x}{2}\right)\cos\left(\frac{2x+\Delta x}{2}\right)}{\Delta x}\,, \tag{218}$$

and

$$\frac{d}{dx}\cos x = \lim_{\Delta x \to 0}\frac{2\sin\left(\frac{2x+\Delta x}{2}\right)\sin\left(\frac{-\Delta x}{2}\right)}{\Delta x}\,. \tag{219}$$

After applying the small angle approximations (Eq.196 and Eq.197) and simplifying, the above equations become,

$$\frac{d}{dx}\sin x = \lim_{\Delta x \to 0} \cos\left(\frac{2x + \Delta x}{2}\right) ,$$

(220)

and

$$\frac{d}{dx}\cos x = \lim_{\Delta x \to 0} -\sin\left(\frac{2x + \Delta x}{2}\right) .$$

(221)

Now we only need to take the limit to obtain,

$$\frac{d}{dx}\sin x = \cos x ,$$

(222)

and

$$\frac{d}{dx}\cos x = -\sin x .$$

(223)

## C.1   Alternative Derivation

Starting with the Fundamental Trigonometric Identity (Eq.6) and taking the derivative of both sides gives,

$$\frac{d}{d\theta}\sin^2\theta + \frac{d}{d\theta}\cos^2\theta = \frac{d}{d\theta}1$$
$$2\sin\theta\frac{d}{d\theta}\sin\theta + 2\cos\theta\frac{d}{d\theta}\cos\theta = 0 ,$$

(224)

where we have used the chain rule of differentiation. Dividing both sides of the above equation by 2 and rearranging gives,

$$\sin\theta\frac{d}{d\theta}\sin\theta = -\cos\theta\frac{d}{d\theta}\cos\theta .$$

(225)

From this we see there are two possible solution sets,

$$\frac{d}{d\theta}\sin\theta = \cos\theta, \tag{226}$$

$$\frac{d}{d\theta}\cos\theta = -\sin\theta, \tag{227}$$

or

$$\frac{d}{d\theta}\sin\theta = -\cos\theta, \tag{228}$$

$$\frac{d}{d\theta}\cos\theta = \sin\theta. \tag{229}$$

From Fig.2, we can see that as we increase $\theta$ in the first quadrant $(0 < \theta < \pi/2)$, $\sin\theta$ (the $y$-component of the circle) increases and $\cos\theta$ (the $x$-component of the circle) decreases. This means that in the first quadrant, we must have $\frac{d}{d\theta}\sin\theta > 0$ and $\frac{d}{d\theta}\cos\theta < 0$. Since $\sin\theta$ and $\cos\theta$ are both positive in this quadrant, the first solution set must be the only one that is correct.

# D  Hyperbolic Trigonometry

We have already seen in Sec.2.1 that ordinary trig functions (sine and cosine) can be defined using a unit circle centered at the origin. For this reason, these functions could be called circular trig functions. We can define the hyperbolic trig functions in a similar way by starting with a hyperbola centered at the origin (see Fig.14). This hyperbola satisfies the equation,

$$\frac{x^2}{a^2} - \frac{y^2}{b^2} = 1 . \tag{230}$$

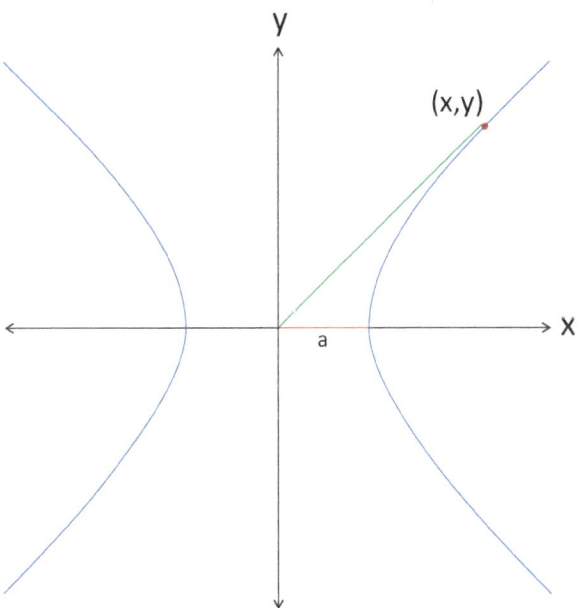

Figure 14: The hyperbola is defined by the equation $\frac{x^2}{a^2} - \frac{y^2}{b^2} = 1$. The foci are located at $x = \pm c$ where $a^2 + b^2 = c^2$.

The functions $\sinh \theta$ and $\cosh \theta$ are defined such that,

$$x = a \cosh \theta \,, \tag{231}$$

$$y = b \sinh \theta \,, \tag{232}$$

where $x$ and $y$ are points on the right side of the hyperbola. Combining these equations with Eq.230 gives the Fundamental Hyperbolic Trigonometric Identity,

$$\cosh^2 \theta - \sinh^2 \theta = 1 \,. \tag{233}$$

The hyperbolic angle, $\theta$, does not represent an angle in the Euclidean sense and should not be confused with a circular angle like we used to describe the points on a circle. The physical interpretation of $\theta$ is that it is related to the area bounded by the $x$-axis, the right side of the hyperbola, and the line containing the origin and a point on the hyperbola. We will derive the exact relationship later but, for now, we can simply regard $\theta$ as a parameter used to determine points on the hyperbola and not worry about a physical interpretation.

## D.1 Derivatives

Taking the derivative of both sides of Eq.233 gives,

$$
\begin{aligned}
\frac{d}{d\theta} \cosh^2 \theta - \frac{d}{d\theta} \sinh^2 \theta &= \frac{d}{d\theta} 1 \\
2 \cosh \theta \frac{d}{d\theta} \cosh \theta - 2 \sinh \theta \frac{d}{d\theta} \sinh \theta &= 0 \,,
\end{aligned}
\tag{234}
$$

where we have used the chain rule of differentiation. Dividing both sides by 2 and rearranging gives,

$$\cosh \theta \frac{d}{d\theta} \cosh \theta = \sinh \theta \frac{d}{d\theta} \sinh \theta \,. \tag{235}$$

From this we see there are two possible solution sets,

$$\frac{d}{d\theta} \sinh \theta = \cosh \theta , \qquad (236)$$

$$\frac{d}{d\theta} \cosh \theta = \sinh \theta , \qquad (237)$$

or

$$\frac{d}{d\theta} \sinh \theta = -\cosh \theta , \qquad (238)$$

$$\frac{d}{d\theta} \cosh \theta = -\sinh \theta . \qquad (239)$$

From Fig.14, we can see that as we increase $\theta$ in the first quadrant $(0 < \theta < \infty)$, $\sinh \theta$ (the $y$-component of the right side of the hyperbola) and $\cosh \theta$ (the $x$-component of the right side of the hyperbola) both increase without bound. This means that in the first quadrant, we must have $\frac{d}{d\theta} \sinh \theta > 0$ and $\frac{d}{d\theta} \cosh \theta > 0$. Since $\sinh \theta$ and $\cosh \theta$ are both positive in this quadrant, the first solution set must be the only one that is correct.

Using the same method we used to derive Euler's Identity in Sec.9, we can derive the Hyperbolic Euler Identity. The result is,

$$e^\theta = \cosh \theta + \sinh \theta . \qquad (240)$$

We will now use this result to provide a physical interpretation of the hyperbolic angle, $\theta$.

## D.2 Physical Interpretation of the Hyperbolic Angle

We start by considering the triangle formed by the line segment connecting the origin and a point on the hyperbola, the line segment connecting this point and the $x$-axis that is parallel to the $y$-axis, and the x-axis (see Fig.15). The area of this triangle is $\frac{1}{2}xy$.

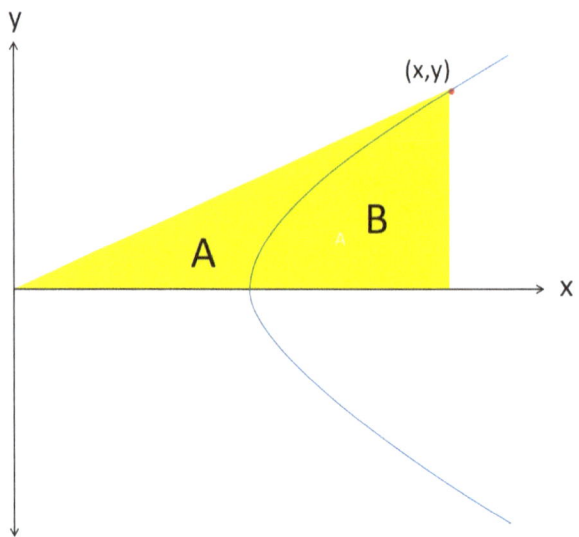

Figure 15: The hyperbolic angle, $\theta$, is related to the area of the portion of the triangle to the left of the curve. This area is calculated by subtracting the area of the portion to the right from the area of the entire triangle. $A = \frac{1}{2}xy - B$.

To calculate the area of the potion of this triangle that is to the right of the hyperbola we need to solve Eq.230 for $y$ and integrate from $a$ to $x$. Solving for $y$ gives,

$$y = b\sqrt{\frac{x^2}{a^2} - 1} \,. \tag{241}$$

Integrating this equation gives,

$$
\begin{aligned}
\int_a^x b\sqrt{\frac{x'^2}{a^2} - 1}\, dx' &= ab \int_1^{\frac{x}{a}} \sqrt{u^2 - 1}\, du \\
&= ab \left[ \frac{u\sqrt{u^2 - 1}}{2} - \frac{1}{2}\ln\left(u + \sqrt{u^2 - 1}\right) \right]_1^{\frac{x}{a}} \\
&= ab \left[ \frac{\frac{x}{a}\sqrt{\frac{x^2}{a^2} - 1}}{2} - \frac{1}{2}\ln\left(\frac{x}{a} + \sqrt{\frac{x^2}{a^2} - 1}\right) \right] \\
&= \frac{xy}{2} - \frac{ab}{2}\ln\left(\frac{x}{a} + \frac{y}{b}\right) \,, \tag{242}
\end{aligned}
$$

where we have used Eq.241 to simplify the fourth line. To get the area of the portion of the triangle that is to the left of the hyperbola we simply subtract the above result from the area of the entire triangle.

$$
\begin{aligned}
area &= \frac{1}{2}xy - \left(\frac{xy}{2} - \frac{ab}{2}\ln\left(\frac{x}{a} + \frac{y}{b}\right)\right) \\
&= \frac{ab}{2}\ln\left(\frac{x}{a} + \frac{y}{b}\right) \\
&= \frac{ab}{2}\ln\left(\cosh\theta + \sinh\theta\right) \\
&= \frac{ab}{2}\ln\left(e^{\theta}\right) \\
&= \frac{ab}{2}\theta \,,
\end{aligned}
\tag{243}
$$

where we have used the definitions of $\cosh\theta$ and $\sinh\theta$ (Eq.231 and Eq.232) in the third line, the Hyperbolic Euler Identity (Eq.240) in the fourth line, and the fact that logarithms and exponentials are inverse functions in the fifth line. The final result is,

$$
\theta = \frac{2A}{ab} \,,
\tag{244}
$$

where $A$ is the area of the portion of the triangle to the left of the hyperbolic curve in Fig.15.

www.ingramcontent.com/pod-product-compliance
Lightning Source LLC
Chambersburg PA
CBHW040840180526
45159CB00001B/262